ECO LIVING MADE SIMPLE

A Practical Guide To Sustainable Choices
For Modern Life

Susan Ribble

Copyright © 2024 Susan Ribble

All rights reserved. No part of this publication may be reproduced, or transmitted in any form or by any means, electronic or mechanical including photocopying, recording or by any information storage and retrieval system without permission in writing from the author.

TABLE OF CONTENTS

INTRODUCTION 6
CHAPTER ONE 10
  Understanding The Basics Of Sustainability 10

CHAPTER TWO 16
  Simple Steps To Reduce Waste 16

  Step One 17

  Step Two 19

  Step Three 21

  Step Four 24

  Step Five 26

  Step Five 28

  Step Six 31

  Step Seven 33

  Step Eight 36

  Step Nine 39

  Step Ten 42

  Step Eleven 45

CHAPTER THREE 48
    Making Your Home Eco Friendly 48

CHAPTER FOUR 56
    Eco-Friendly Clothing And Lifestyle Options 56

CHAPTER FIVE 64
    Benefits of Green Energy 66

CHAPTER SIX 72
    Eco-Friendly Travel And Transportation 72
    Greener Options for Transportation 73
    Tips for Eco-Friendly Travel 75

CHAPTER SEVEN 78
    The Importance Of Ethical Consumption 78
    The Significance of Ethical Consumption 79
    The Role of Consumers in Driving Change 81

CHAPTER EIGHT 84
    Long-Term Habits for Sustainable Living 84

CONCLUSION 90

INTRODUCTION

Have you ever ponder about the size of our environmental footprint? It's the measure of the impact our choices have on our planet. From the food we eat to the way we travel, every decision lives an imprint and it's not just about machines but about water usage, waste production and more. We all have a responsibility to trade lightly.

There has never been a more pressing need for sustainable living in the modern world. Making decisions that lessen our influence on the environment is crucial as environmental degradation and climate change become more urgent problems. However, navigating the overwhelming amount of eco-friendly information can be daunting. That's where ECO LIVING MADE SIMPLE comes in.

In Eco Living Made Simple, you'll discover that sustainability isn't about perfection or making drastic, life-altering changes. It's about adopting small, practical habits that fit seamlessly into your everyday life. These little steps, when multiplied by millions of people, become powerful forces for change. Whether you're taking your first step into eco-conscious living or looking to deepen your impact, this book will guide you through easy, actionable ways to live greener.

You can make changes in your life without having to overhaul everything. Actually, one of the best ways to transition to an eco-friendly lifestyle is to start small with small, manageable changes. You can make changes in your life without having to overhaul everything. Actually, one of the best ways to transition to an eco-friendly lifestyle is to start small with small, manageable changes. These small changes, like using less plastic, using less water, or switching to energy-efficient light bulbs, may make a big difference over time.

One reusable water bottle, for example, might not seem like much, but if millions of people make this tiny shift, we can significantly reduce the amount of plastic garbage that ends up in the environment. These little deeds add up to something big, which is what gives them their power. All that's required is consistency; perfection is not necessary. That's the beauty of eco-living—it's about progress, not perfection.

It's surprisingly simple to start living sustainably. Finding one or two areas in your life where you can make little adjustments is the first step. Perhaps it's setting up a compost pile or selecting reusable shopping bags. Perhaps all it takes is turning off electronics when not in use or switching to plant-based eating a few days a week.

This journey is uniquely yours, and there's no right or wrong way to begin. What's important is finding what resonates with you and making it a habit. Throughout this book, you'll find practical tips that fit into your busy, modern life—so you can begin your journey toward eco-living without feeling overwhelmed.

Sustainability is often seen as something that requires sacrificing convenience or luxury, but that's far from the truth. In today's world, living green can be modern, innovative, and adaptable. Thanks to advances in technology, sustainable options are becoming more accessible and practical, whether it's smart home energy systems, eco-friendly fashion brands, or even electric vehicles. You no longer have to choose between convenience and eco-consciousness.

A modern approach to living green is about integrating sustainable choices into the way we live today—without having to retreat into extreme simplicity or give up the comforts of modern life. From reducing energy consumption to choosing eco-friendly products, it's about finding balance and making informed decisions that align with your values and lifestyle.

The truth is, living sustainably doesn't need to be complicated, expensive, or inconvenient. It's about making mindful choices that accumulate over time. With every small change you make, you're contributing to a larger movement—a movement toward a healthier planet and a more sustainable future.

In Eco Living Made Simple, you'll learn how to make these changes easily and effectively. From your home and diet to how you shop and travel, this guide will show you how simple it can be to live in harmony with the planet, while still enjoying the conveniences of modern life.

Imagine living a more eco-conscious lifestyle without sacrificing comfort or convenience.

Welcome to a simpler, greener way of living.

CHAPTER ONE

Understanding The Basics Of Sustainability

When people hear the word sustainability, they frequently associate it with environmentally friendly practices like recycling. Sustainability is very much more than the environment alone. It simply means the ability of something to be sustained or to last.

The ability of our society to continue and grow without exhausting all of the natural resources required for future generations to live is known as sustainability. Sustainable development works toward this long-term objective by putting frameworks, procedures, and support in place that come from international, national, and local organizations.

Its foundation is a comprehensive strategy that combines environmental, social, and economic aspects to build robust systems that can endure throughout time. The relevance of sustainability in today's society has increased due to rising awareness of social inequality, environmental deterioration, and the limited nature of natural resources.

Principles of Sustainability

Sustainability is fundamentally based on a number of important ideas. Interdependence, which recognizes the complex interactions between people, the environment, and the economy, is one of the most important. Decisions taken in one domain can have a big effect on the others. This interdependence stresses the necessity for a full grasp of how human activities effect the greater environment

Equitable access to resources and opportunities for all people, irrespective of socioeconomic level, background, or geographic location, is a fundamental value. Social justice and fairness must be promoted by sustainable practices so that everyone has an opportunity to prosper. A further essential component of sustainability is conservation. In order to stop depletion and safeguard biodiversity, it entails the prudent use and management of natural resources. This covers activities like waste minimization, recycling, and sustainable material procurement. Furthermore, sustainability depends on systems' resilience—their ability to tolerate shocks and changes and adjust accordingly. Resilient systems are able to bounce back from setbacks and carry on with their work, which emphasizes the need of long-term planning and vision in sustainability initiatives.

Dimensions of Sustainability

Three primary categories may be used to group the characteristics of sustainability: social, economic, and environmental sustainability. Environmental sustainability focuses on managing natural resources sustainably, preserving ecological balance, and tackling climate change. It includes programs to increase energy efficiency, support renewable energy sources, and lower greenhouse gas emissions.

In contrast, economic sustainability entails implementing sustainable company strategies that reduce environmental damage and encourage social responsibility. This might involve cutting waste, making sure that fair labour standards are followed, and utilizing sustainable resources. In order to reduce environmental dangers and promote economic growth, the shift to a green economy is essential.

The concept of social sustainability highlights the significance of cultural preservation and community involvement. Sustainable practices should engage local people in decision-making processes, ensuring their views are heard and their needs satisfied. This component also includes upholding indigenous rights, preserving local customs, and safeguarding cultural variety and legacy.

The Role of Individuals

Individuals' everyday decisions and actions have a significant impact on the advancement of sustainability. Resource conservation and waste reduction may be greatly increased by using the reusing, recycling, and reducing strategies. Using public transportation, bicycling, walking, carpooling, or other sustainable modes of transportation can help cut down on carbon emissions and enhance air quality. Furthermore, the desire for sustainable practices is fuelled by conscious consumerism, which includes buying eco-friendly items, supporting small, sustainable enterprises, and being aware of one's own consumption habits. Both advocacy and education are crucial; spreading knowledge of environmental protection laws and sustainability-related concerns can result in more significant changes in society.

Challenges to Sustainability

There are still many obstacles in the way of sustainable practices, even with increased knowledge and adoption of them. Economic obstacles frequently obstruct the shift to sustainability since it may call for an initial financial outlay that people and companies cannot afford. Sustainable products might often be seen as more expensive, restricting their accessibility.

Effective sustainability projects require political will, and a lack of support from the government might impede these efforts. Moreover, cultural opposition to change can make things more difficult since it puts established customs and ways of life in jeopardy. In order to remove these obstacles, community involvement and education are needed to emphasize the advantages of sustainable practices.

Global inequality also presents serious obstacles to sustainability. Different nations have different requirements and capacities; underdeveloped countries frequently have particular difficulties like poverty and poor infrastructure, which makes it more difficult for them to adopt sustainable practices.

Governments, corporations, and people all need to grasp the fundamentals of sustainability. Through acknowledging the interdependencies among social justice, economic stability, and environmental health, stakeholders may collaborate to establish sustainable systems that will benefit both present and future generations. Adopting sustainability is not only a duty but also a chance to promote creativity, improve people's quality of life, and guarantee a healthy world for future generations. A common goal of a more just and resilient world must guide community action and dedication on the path toward sustainability.

CHAPTER TWO

Simple Steps To Reduce Waste

Numerous choices we make in our daily lives have a significant effect on the environment.

In an age where consumption is often prioritized over sustainability, reducing waste has become an essential practice for individuals and communities alike. Every day, countless items are discarded, contributing to overflowing landfills and environmental degradation. This growing waste crisis not only harms the planet but also affects human health and resource availability.

By taking proactive measures to decrease waste, people may empower themselves to make significant changes in their everyday lives. Through a collective re-evaluation of our purchasing habits and adoption of sustainable behaviours, we may reduce our environmental footprint and foster a more robust ecosystem. Every tiny action advances a greater cause of sustainability by encouraging accountability and community involvement.

Realizing that every action matters is crucial as we set out on this quest to reduce waste. We may make a positive impact on the world that benefits present and future generations by adopting these small changes into our daily lives.

Step One

Practice Mindful Consumption

The essential practice of mindful consumerism encourages people to make thoughtful decisions about what to purchase. In a culture where marketing and consumerism are rampant, stopping to examine if a purchase is really essential can lead to a more sustainable way of life and less waste.

Before you acquire something, consider your options and determine whether you truly need it. Consider its objective and if it will make a difference in your life. It's important to distinguish between necessities and luxury things, and to consider whether you currently own something similar that serves the same purpose. Think about whether you will actually need the item in the long term or whether it is simply a whim.

Impulsive purchases might lead to the collection of items you don't actually need. Try establishing a waiting period before making non-essential purchases to remedy this. Giving yourself a whole day to consider the purchase, for instance, could assist you in determining if it was an emotional or necessary expenditure.

After you've determined that you must buy anything, give it some thought. Look for products that align with your values, such as those that come from sustainable or ethical manufacturing. By comparing prices, brands, and materials, you can support environmentally conscious businesses and promote a more sustainable economy by making informed selections.

Assessing quality as opposed to quantity is only one more crucial aspect of conscious consumption. Investing in durable, well-made, and high-quality items can reduce the frequency with which they need to be replaced. Certain things may be more expensive initially, but they often save money over time and produce less waste. Seek for items with traditional designs, extended lifespan, and reparability.

A minimalist mindset might simplify and eliminate clutter in life. Strive to keep only helpful or joyful items in your possession. This tactic not only reduces consumption but also promotes a tidier and organized living space.

You could shift your focus from wanting more to appreciating what you already have. Regular reflection on the things you love and the experiences they provide, rather than obsessing on what you lack, can help reduce the urge for unnecessary purchases.

Mindful consumerism is based on making deliberate choices that align with your values and requirements. A more sustainable lifestyle and a significant reduction in waste may be achieved by carefully weighing your options, avoiding hasty purchases, and emphasizing quality over quantity. This tactic encourages you to be more grateful for what you own while simultaneously protecting the environment, leading to a happier and more fulfilling life.

Step Two

Opt for Reusable Items

Replacing single-use goods with reusable alternatives is one of the most significant actions people can take in the fight against waste. The growth of throwaway items, such bottles, containers, and bags made of plastic, has greatly added to the expanding waste problem. Reusable products allow us to reduce our environmental impact and promote sustainable practices that improve our way of life and the environment.

When you go shopping, switching to cloth bags from plastic ones is an easy yet powerful adjustment. Nowadays, a number of retailers have strong, machine-washable cotton bags that are made to last and can accommodate many goods. You can lessen the need for single-use plastics and help lessen the amount of plastic waste that ends up in landfills and the ocean by packing your own bags.

Glass or stainless steel drinking bottles offer a robust and environmentally friendly substitute for single-use plastic bottles. These reusable bottles not only reduce the need for single-use plastic, but they also keep drinks cold and fresh. Reusable bottle use promotes staying hydrated and helps cut down on the quantity of plastic bottles that wind up in landfills every year.

Purchasing reusable food storage containers is crucial to reducing waste in this area. Meal prep, leftovers, and snacks can be stored in glass or BPA-free plastic containers. These containers reduce food waste by helping to preserve food freshness in addition to being environmentally beneficial. Reusable beeswax wraps can also reduce the amount of single-use plastic consumed by wrapping food in them rather than using plastic wrap.

Using reusable products promotes sustainability and accountability in addition to ease of use. We can have a major influence on waste reduction by implementing simple, deliberate adjustments in our everyday activities. Making the switch to reusable alternatives helps the environment and encourages others to do the same, starting a movement toward a more sustainable future. By these initiatives, we can improve the health of the environment and encourage actions that protect resources for future generations.

Step Three

Buy in Bulk

Purchasing in large quantities is an effective way to promote sustainable consumption and cut waste.

Considering how common single-use packaging is becoming, buying goods in bulk can help reduce the quantity of plastic and cardboard that often ends up in landfills. Customers who choose to buy in bulk can save money and lessen their environmental effect while also adopting a proactive attitude toward their purchasing practices.

One of the key advantages of buying in bulk is the decrease in packing waste. These days, a lot of grocery stores sell a wide range of products in bulk bins, such as cereals, nuts, seeds, dried fruits, and even certain cleaning goods. These bulk bins may be used to fill your own reusable containers or bags, which essentially eliminate the need for more plastic. This promotes more attentive consumption practices in addition to lowering the trash produced by single-use packaging.

Purchasing in bulk frequently results in financial savings as well as a reduction in packing waste. Generally speaking, bulk purchases are less expensive than packaged goods, which helps customers save money over time. With their lengthy shelf lives, staples like rice, pasta, and lentils benefit most from this. By shopping in bigger numbers, you may stock up on staples while spending less per unit.

Additionally, making larger purchases may result in fewer journeys to the store, which lessens the carbon footprint caused by transportation. You may save time and energy and help reduce greenhouse gas emissions by making fewer journeys to the supermarket.

It is important to prepare ahead of time when opting to purchase in bulk. List the things you use most often and figure out how much you really need to prevent waste. Although buying in bulk is a sustainable option, be sure you can finish the products before they go bad or stale.

Purchasing in bulk is a sensible and efficient approach to encourage sustainable consumption and lessen packaging waste. You may save money, be more convenient, and lessen your environmental effect by choosing retailers that sell bulk bins and filling your own reusable containers. This small change in your shopping habits benefits you and the greater movement to reduce waste and promote a healthy world.

Step Four

Plan Meals and Reduce Food Waste

One of the most important tactics for minimizing food waste and encouraging sustainability in the kitchen is meal planning. Making a well-thought-out meal plan helps people buy less food overall, make sure items are used effectively, and reduce the chance of spoiling. Because less food gets up in landfills, this proactive strategy not only helps save money but also promotes a healthy environment.

Taking stock of what you already have in your cupboard, fridge, and freezer is the first step in meal planning. By evaluating your current items, you may include them in your meal plan and lower the likelihood that they will go to waste. You may make a weekly meal plan that specifies what you will consume once you have a firm grasp on your inventory. The strategy ought to take into account the expiration dates of perishable goods, giving priority to those that require immediate usage.

When establishing your food plan, aim for diversity and balance. To make well-rounded meals, incorporate a variety of cereals, fruits, vegetables, and meats. Create a thorough shopping list based on your strategy to make sure you only buy what you need and avoid overspending and impulsive purchases.

Proper storage of leftovers is a crucial component in cutting down on food waste. To keep food fresh, store it in airtight containers and label it with dates so you know exactly what has to be eaten first. There are many of inventive ways to recycle leftovers so they become fresh dinners or snacks. For instance, you may combine fruits into smoothies, add roasted veggies to salads, and include grains into soups. Thinking creatively allows you to.

In addition, think about storing food according to the "first in, first out" principle. This entails arranging more recent products behind more ancient ones so that the latter get depleted first. This easy procedure helps reduce waste and spoiling.

Meal planning and efficient food storage management are two important strategies for cutting down on food waste. People may help create a more sustainable kitchen environment by planning thoughtful meals, coming up with inventive ways to use leftovers, and using the right storage methods. This leads to a healthier connection with what we eat and how we consume it, as well as a deeper appreciation for food and its place in our lives. It also saves money.

Step Five

Learn About Recycling

Recycling is essential to both environmental sustainability and trash reduction. But efficient recycling necessitates a thorough comprehension of regional regulations and appropriate sorting techniques. Understanding the local recycling regulations can help you make sure that recyclables are handled properly and eventually make the world a healthier place.

Recycling regulations differ greatly from town to community. It's critical to learn about the items that your local program accepts for recycling. Paper, cardboard, glass, and certain plastics are frequently recyclable materials, although not all materials qualify. Many recycling centers have strict guidelines on the kinds and states of items they may take. For example, particular recycling codes may need to be applied to various plastics, and containers that have held food, for example, may need to be cleaned before recycling. It will be easier for you to decide what may and cannot be recycled if you are familiar with these rules.

Proper sorting becomes crucial after you know your local recycling regulations inside and out. Set aside certain bins or containers in your house for various recyclable products. Keeping things in their proper places makes recycling easier and also reduces the risk of contamination that can happen when recyclables and non-recyclables are combined. One serious problem that can result in recycling effort.

Avoiding "wish-cycling," which is the habit of putting non-recyclable things in the recycling bin in the hopes that they would be recycled, is another crucial part of recycling. Wish-cycling eventually impedes the recycling process and presents difficulties for recycling facilities. Check the approved products list for your location frequently and make sure you follow the rules to avoid this.

An essential component of efficient recycling is education. To help locals understand the recycling process, several municipalities provide workshops, online courses, or instructional leaflets. Participating in these instructional programs can improve your understanding and provide you the tools you need to become a more efficient recycler.

Reducing waste and increasing sustainability begin with education on recycling. You may make sure that recyclable items are processed correctly by being informed with your local recycling regulations, separating your garbage appropriately, and refraining from wish casing. This dedication not only helps create a more sustainable environment, but it also sets an example for others and promotes waste management and responsible consumption in the community. By using responsible recycling techniques, everyone of us can contribute to resource conservation and preserving our world for coming generations.

Step Five

Start Composting

Composting is a revolutionary method that improves the soil and decreases organic waste, making it a crucial first step toward sustainable living. Composting yard trash, coffee grounds, and fruit and vegetable peels is one way that people may drastically reduce the quantity of organic material that ends up in landfills, which releases greenhouse gases. Instead of tossing these materials away, composting allows you to repurpose them into a beneficial resource for gardening and landscaping.

Organic debris is broken down throughout the composting process via spontaneous decomposition. Worms, microbes, and other creatures that live in compost piles assist in this process. The end product is nutrient-rich compost, which may be applied to gardens to increase water retention, plant development, and soil quality. Composting promotes sustainable farming methods and a healthy environment by reintroducing these nutrients to the soil.

The process of beginning composting is not too complicated. A variety of composting techniques are available, such as conventional outdoor compost bins, vermin-composting (with worms), and compact interior composting systems for spaces that are limited. The secret is to choose a technique that works for both your gardening demands and living circumstances.

To construct a good compost pile, it's vital to balance "greens" (nitrogen-rich materials) with "browns" (carbon-rich materials). Greens are leftovers from the kitchen, such as fruit and vegetable peels, coffee grounds, and grass clippings; browns include things like cardboard, dry leaves, and twigs. This equilibrium produces high-grade compost and aids in the decomposition process.

For composting to be successful, proper aeration and moisture levels are also essential. Keep your compost wet enough to encourage microbial activity and turn it frequently to add oxygen, which promotes decomposition. You may add water to your compost if it seems too dry, and you can add more browns to your compost if it seems too wet to help balance the moisture level.

Seeing kitchen scraps and yard trash gradually turn into rich compost is one of the most satisfying parts of composting. This final product offers a sustainable substitute for chemical fertilizers and can be used to improve lawns, potted plants, and garden beds.

An important first step in lowering organic waste and advancing sustainability is to start a composting habit. Composting yard waste and kitchen scraps helps people reduce their environmental footprint while improving the soil's health and the plants' vibrancy. This easy-to-do but efficient method allows us to take care of the environment and reap the rewards of our labours at the same time. It combines the ideas of recycling and conservation. By composting, we may develop a more environmentally friendly way of living, encourage others to do the same, and strengthen our bond as a community.

Step Six

Fix Rather than Replace

The desire to replace broken or damaged products frequently outweighs the possible advantages of repair in a consumer-driven society. Adopting the mentality of fixing rather than replacing something not only increases the lifespan of your possessions but also greatly lowers waste and encourages sustainability. By encouraging a more deliberate connection with the things we own, this strategy promotes environmental responsibility, creativity, and resourcefulness.

Whether it is furniture, appliances, or clothes, the first impulse could be to throw it away and get a new one when something breaks. However, many products can be mended with a little work and ingenuity. For instance, small rips in garments may be sewn shut, and missing buttons are simple to replace. Clothes that need minor repairs can be given new life so you can wear them for years to come rather than throwing them in the trash.

Many common problems with appliances and electronics can be resolved by simple troubleshooting or inexpensive part replacement. By helping you through the repair process, online tutorials and tools can enable you to take on tasks you may have previously believed were outside the scope of your expertise. Furthermore, a few towns have repair workshops where you can work with knowledgeable people to learn how to mend different things. These events give you useful skills and promote a feeling of community.

Additionally, furniture offers a great chance to fix rather than replace. With a little imagination and elbow grease, worn-out upholstery, scuffed surfaces, and broken chair legs may often be restored or refinished. Refinishing, painting, and reupholstering furniture are examples of do-it-yourself tasks that can turn it into distinctive pieces that suit your style and cut down on waste.

In the long term, we save money when we repair goods in addition to keeping garbage out of landfills. Repairing something can frequently be less expensive than buying a new one, especially for high-quality items that may be pricey to replace. Additionally, fixing things encourages a sustainable culture in which we place a higher priority on durability and quality than on throwaway goods.

In the end, a change in viewpoint is necessary to adopt a repair mindset. Broken objects can be seen as chances for innovation and problem-solving, as opposed to being seen as waste. This adjustment not only lessens our influence on the environment but also fosters a greater appreciation for the possessions we have.

One effective strategy for reducing waste and increasing sustainability is to repair rather than replace. By embracing this practice, individuals can extend the life of their goods, minimize waste, and build a more thoughtful relationship with consuming. In addition to being good for the environment, fixing things fosters ingenuity and resourcefulness, which will help create a more sustainable future for present and future generations.

Step Seven

Donating and up cycling

Donating and up cycling are two effective methods that assist prolong the life of things we no longer need in a world where sustainability is becoming more and more important. Both strategies decrease waste and encourage innovation and community involvement, which makes them vital elements of a sustainable way of living.

Up cycling is the process of creatively repurposing objects that might otherwise be thrown away to create new, useful goods. This method pushes you to recognize potential in objects that you might normally consider rubbish, which fosters creativity and resourcefulness. For example, worn-out clothing can be made into quilts, bags, or home decor items; old furniture can be refinished or recycled into new designs; and glass jars can be turned into elegant vases or storage containers. Your creativity is the only restriction when it comes to up cycling, which makes it an enjoyable and fulfilling endeavour.

Up-cycling preserves resources and lessens its negative effects on the environment by keeping garbage out of landfills and lowering the need for new products. You may make one-of-a-kind, customized pieces that embrace sustainable techniques and showcase your flair by repurposing outdated objects.

However, giving away goods that are still in good condition is an additional efficient method of cutting waste and improving your neighbourhood. Many people own things that are in good shape but have lost their meaning in their lives. Consider giving these things to neighbourhood shelters, charities, or community organizations rather than throwing them out. This method guarantees that reusable things have an opportunity to be reused and valued by others in addition to helping people in need.

Donation centres sometimes take in a variety of commodities, such as toys, furniture, household goods, and clothes. Make sure your things satisfy the requirements of the organization by reading their rules before making a donation. By giving, you may help lessen the pressure on landfills and encourage a culture of sharing and sustainability within your community.

By creatively transforming items you no longer use or sharing them with those in need, you contribute to a healthier environment and a stronger community. Embracing these practices not only benefits the planet but also enriches your life by fostering creativity and connection with others. Up cycling can inspire creativity and craftsmanship, while donating encourages community support and engagement. These practices reflect a shift away from a throwaway culture toward one that values sustainability and resourcefulness.

Step Eight

Cut Down on Paper Waste

One of the most important steps on the path to sustainability and environmental preservation is cutting down on paper waste. There are several ways to reduce our dependency on paper goods in an increasingly digital environment, which will help to cut down on the quantity of waste paper use produces. We may promote more sustainable practices in our daily lives and help the environment by switching to digital alternatives and using less paper.

Making the switch to digital is one of the best strategies to cut down on paper waste. Nowadays, a lot of businesses use e-billing services, which let customers get their bills via email rather than paper statements. This small change helps to simplify money administration by doing away with the requirement for printed bills. Digital receipts and online transactions allow you to reduce paper waste even further and maintain your data in an easily accessible format.

Think about using digital subscriptions for newspapers, magazines, and other media in addition to e-billing. The same material may now be found in digital editions of many print periodicals, which eliminate the need for physical printing and shipping. It is possible to support content creators while cutting back on paper use by subscribing to digital versions.

Using digital note-taking tools can also drastically cut paper usage. Apps like Evernote, Microsoft OneNote, or Notion let you arrange your ideas, tasks, and thoughts digitally rather than writing notes down on paper. These apps frequently have collaboration options, tags, and reminders, among other features that increase productivity. Making the switch to digital note-taking can help you organize your workflow more efficiently and cut down on paper waste.

Furthermore, think about putting waste-reduction strategies into place when using paper is inevitable. When feasible, use both sides of the paper, and select products made of recycled paper. If printing is required, change the parameters to print in draft mode or minimize the margins to conserve paper.

To further reduce paper usage, configure your printer to print multiple pages on a single sheet.

You may encourage a sustainable culture in your community by educating yourself and others on the value of cutting down on paper waste. Encourage loved ones to adopt digital substitutes and spread the word about practical ways to reduce paper usage.

Promoting sustainability and environmental protection need reducing paper waste. People can drastically reduce their dependence on paper goods by choosing digital alternatives like note-taking applications, digital subscriptions, and e-billing. These small but powerful adjustments simplify our daily lives, cut down on waste, and improve the accessibility and organization of information. By adopting these behaviours, we promote a society that values conservation and responsible consumption and help create a more sustainable future.

Step Nine

Take Part in Community Projects

Participating in community projects is a potent method to strengthen waste minimization activities and promote a shared commitment to sustainability. Participating in neighbourhood clean-ups, recycling campaigns, and swap meets allows people to meet people who have similar beliefs and aspirations while also improving the community and the environment. Participation like this promotes awareness of the value of conserving natural resources and cutting waste while strengthening ties within the community.

Local clean-up campaigns offer a fantastic way to leave a lasting impression on your neighbourhood. These gatherings usually concentrate on tidying up public areas where trash and rubbish can gather, such parks, beaches, or neighbourhoods. Taking part in a clean-up not only makes the region more beautiful, but it also increases awareness of how important it is to maintain a clean environment. It acts as a concrete reminder of our shared accountability for preserving the quality of our surroundings.

Another great approach to interact with your community and promote sustainable practices is through recycling drives. To encourage citizens to recycle things that might not be accepted through routine curbside pickup, many cities host recurring recycling events. Textiles, other specialist materials, and electronic garbage are frequently collected during these drives. You can assist in keeping more rubbish out of landfills and educate people about appropriate recycling techniques and the value of waste reduction by taking part in these events.

Community swap meets offer an innovative and environmentally friendly method of repurposing goods while promoting a sense of belonging. People can bring things they no longer need to these events and trade them for stuff brought by others. Giving things a second chance at life not only helps reduce trash but also promotes a sharing and reuse culture within the community. You can find new treasures, reduce trash, and save money by taking part in swap meetings.

Additionally, you can increase your influence by joining advocacy or environmental groups in your community. These groups frequently hold campaigns, workshops, and events with an emphasis on conservation, sustainability, and waste minimization. Forming alliances with people who share your values can help you be more dedicated to environmental stewardship and will provide you access to more information and tools to help change the world.

Taking part in community projects also helps you develop a sense of pride and accountability for the environment. When individuals band together for a similar goal, they form a network of support that multiplies their efforts and inspires others to follow suit. By working together, we may create long-lasting modifications to the customs and behaviours of the society, establishing sustainability as a shared value.

One of the most important ways to support sustainability and waste reduction effort is to participate in community projects. People can engage with their neighbours, spread knowledge of environmental issues, and take concrete actions toward a cleaner, healthier neighbourhood by organizing clean-up events, recycling drives, and swap meets. By taking part in these events, we may encourage one another and foster a sustainable culture, cooperating to build a better future for present and future generations.

Step Ten

Teach Others and Yourself.

Promoting a more sustainable lifestyle and a healthier environment begins with educating yourself and others about waste reduction techniques. By keeping up with the most recent waste management techniques and trends, you give yourself the power to make decisions that will have the least negative effects on the environment. Furthermore, imparting your knowledge to loved ones, friends, and the larger community can have a beneficial knock-on impact that promotes group action in the direction of sustainability.

The search for trustworthy resources of waste reduction information is the initial stage in this educational process. There are a plethora of websites, books, and organizations that concentrate on composting, recycling, sustainability, and mindful consumption. You can gain a deeper awareness of the waste-related issues we face and the range of options available to address them by looking through these materials. Attending community classes, webinars, or workshops can help you learn more and meet people who share your enthusiasm for sustainability while also advancing your knowledge.

After gathering data, think about imparting your knowledge to others. Talking with friends and family about trash reduction techniques can start important discussions about the value of sustainability and the effects of individual actions. You may offer advice on how to start a compost bin, plan recycling initiatives, or use less plastic. You can encourage people around you to take up similar routines and behaviours by encouraging an open discussion about these subjects.

Creating educational materials, such as brochures or info graphics, can be an efficient approach to distribute information. These resources can provide an overview of important waste reduction techniques, regional recycling policies, or the advantages of composting, making it simple for others to comprehend and incorporate these practices into their everyday life. Another option is to arrange casual get-togethers or workshops where you may impart your knowledge and offer practical examples.

Social media sites provide an additional means of raising awareness on trash minimization. Through disseminating articles, information, and personal accounts, you can expand your reach and inspire others to consider their purchasing patterns. You can also reach out to others who are enthusiastic about changing the world and spread the word about your message by interacting with local sustainability groups on the internet.

Furthermore, one of the best methods to educate people is to set a good example for others. Your friends and family are more inclined to think about implementing similar habits when they see your dedication to minimizing waste, whether it is through recycling, composting, or using reusable items. Your deeds can act as a beacon of hope and encouragement for others, proving that living sustainably is both rewarding and feasible.

A crucial first step in creating a sustainable future is educating yourself and others about waste reduction. You can start a beneficial domino effect that inspires people and communities to embrace environmentally friendly practices by remaining informed and disseminating your knowledge. By working together, we can bring about significant improvements in waste management, environmental stewardship, and consumption trends. By working together, we can foster a sustainable culture that benefits the environment and the next generation.

Step Eleven

Establish Waste Reduction Objectives

Establishing targets for waste reduction is a crucial tactic for encouraging environmentally conscious behaviour and sustainable habits. People can develop a systematic strategy for reducing waste in their daily lives by setting clear, attainable goals. In addition to providing a course of action, these objectives provide you a sense of success as you monitor your advancement, which motivates you to keep up your commitment to sustainability.

It is crucial to start small and manageable at first. For example, you may set a monthly goal to cut back on plastic use by a specific proportion. This could be avoiding products with excessive plastic packaging, using a glass or stainless steel water bottle instead of purchasing bottled water, or choosing to use reusable bags instead of single-use plastic bags. You may make significant improvements without feeling overburdened by concentrating on a small number of important topics.

Composting is yet another fantastic objective to think about. Making the commitment to begin composting if you do not already can help cut down on the amount of organic waste that ends up in landfills. Start by using a countertop composting device or putting up a basic compost bin in your backyard. Keep a record of how much of your cooking scraps you get out of the garbage, and see how your compost improves your garden over time. This is beneficial to the environment and offers a satisfying way to gauge the success of your efforts.

Monitoring your progress is essential if you want to make sure you stay motivated. To document your accomplishments and any difficulties you face along the road, think about maintaining a notebook or downloading an app. By reflecting on your progress and marking significant anniversaries, this practice helps you to strengthen your resolve to reduce waste. Setting deadlines for your goals, like monthly or quarterly check-ins, may help you determine what is working and where revisions may be necessary.

You can progressively increase your objectives as you become more assured of your capacity to cut waste. Once you have successfully cut back on plastic, for instance, you could aim to completely eradicate food waste by organizing meals more efficiently or coming up with inventive methods to repurpose leftovers. By building on your past successes with each new objective, you may create a sustainable lifestyle that changes with time.

Think about communicating your goals to friends and family in addition to your own aspirations. As a result, everyone may be inspired to follow comparable waste reduction initiatives within a network of support. Even better, you might create team challenges or take part in community projects that support your objectives to double the effect of your work and strengthen the bonds within the community.

Setting waste reduction objectives is a great strategy to create sustainable behaviours and foster a commitment to environmental care. You may make a significant difference in your life and the life of your community by setting modest but attainable goals, monitoring your progress, and making necessary adjustments to your original plan. In addition to reducing trash, this trip encourages others to follow suit, building a movement toward a more sustainable future. One goal at a time, working together, we can make a difference.

CHAPTER THREE

Making Your Home Eco Friendly

One of the most important steps toward sustainability and environmental stewardship is making your house eco-friendly. You may greatly lower your carbon footprint and contribute to a healthier planet by making wise decisions and practices in your living area. Making your home more sustainable needs a holistic strategy that encompasses energy efficiency, waste reduction, sustainable materials, and mindful consumption.

Boost Your Energy Efficiency

The first step in lowering the carbon footprint of your house is energy efficiency. Energy-efficient LED lighting, which consumes substantially less energy and has a far longer lifespan than incandescent bulbs, should be used as a starting point. Invest in appliances with an Energy Star rating; these units are made to use less energy while maintaining the same level of performance as regular appliances.

Installing a programmable thermostat, which enables you to set different temperatures at different times of the day, can help your heating and cooling systems operate at their best. When you are not at home, this feature makes sure that energy is not wasted. Proper insulation is also crucial; it reduces heat loss in winter and keeps your home cool in summer. Check for drafts around windows and doors in your house, and use caulk or weather stripping to seal any openings. Over time, these modest adjustments can result in considerable energy savings and decreased utility costs.

Adding renewable energy sources, such solar panels, can improve energy efficiency even more. Even while the upfront cost could be high, it might be justified in the long run due to the reduction in energy costs and the advantages it has for the environment. Homeowners can now more easily afford to install solar energy systems thanks to numerous regions' incentives and rebates.

Save Water

Another essential component of creating an eco-friendly home is water conservation. Installing low-flow showerheads, toilets, and faucets is a good place to start since they save water without compromising functionality.

With a 50% reduction in water usage, these fixtures will have a major effect on your water cost.

Inspect your plumbing on a regular basis for leaks, and address them right away. Even tiny leaks over time could cause a large amount of water waste. Additionally, you can save water by using the dishwasher or washing machine only when there are full loads, taking shorter showers, and shutting off the faucet while brushing your teeth.

To gather rainwater from your roof, think about installing a rainwater collecting system. Reliance on municipal water supplies can be decreased by using this collected water for outdoor surface washing, gardening, and irrigation. By using these water-saving techniques, you can reduce your water utility bills while simultaneously protecting this valuable resource.

Make Use of Sustainable Materials

Using sustainable materials is essential when remodelling or redecorating your house. Choose things that are renewable, recyclable, or sustainably sourced. For example, bamboo is a great option for flooring and a resource that replenishes quickly. Reclaimed wood, which is taken from dilapidated structures or furnishings, not only minimizes waste but also gives your room a distinctive charm.

Recycled materials, such as composite materials created from post-consumer products or recycled glass, should be taken into consideration when choosing countertops. Additionally, to improve indoor air quality, use low-VOC (volatile organic compounds) paints and finishes, which release fewer dangerous chemicals into the air. Your health and the environment both gain from this decision since it limits your exposure to harmful pollutants.

Using sustainable materials improves the quality of your living space and reduces your environmental impact. By making informed choices about the resources you use, you can contribute to a more sustainable economy and encourage responsible production practices.

Reduce Waste Effective waste reduction methods are crucial for constructing an eco-friendly home. Put in place a thorough recycling system first. To make sure you sort recyclables appropriately, familiarize yourself with your local recycling regulations. Make it simple for your family to recycle by designating bins for various recyclable materials, such as paper, plastics, glass, and metals.

Composting is an additional effective method of waste reduction. Use a countertop composting system or establish a compost bin in your backyard for kitchen trash, such as coffee grinds, fruit and vegetable peels, and yard garbage. In addition to keeping organic waste out of landfills, composting produces nutrient-rich soil that can improve your garden.

You can further cut waste in your home by promoting mindful consumption. Opt for minimally packaged products, encourage purchase in bulk, and prioritize reusable things over single-use ones. Teaching your family the value of waste reduction can help foster a sustainable culture at home by encouraging everyone to make more thoughtful consumption choices.

Select Eco-Friendly Items

Choosing eco-friendly cleaning and personal care products is a crucial part of keeping your house green. Conventional cleaning products frequently include dangerous chemicals that can have an adverse effect on indoor air quality and be hazardous to human health. Rather, use environmentally friendly, non-toxic, biodegradable cleaning products that are efficient without harming the environment.

Think about creating your own cleaning products at home with basic components like baking soda, vinegar, and essential oils. Without using dangerous chemicals, these natural substitutes for commercial cleaners can be just as effective. In a similar vein, seek out businesses that emphasize natural ingredients and eco-friendly packaging when selecting personal care products. These days, a lot of businesses sell organic, cruelty-free, and resource-renewable items.

By making eco-friendly product choices, you promote environmentally conscious businesses and practices while also preserving the health of your family. This conscientious consumption stimulates the market to move toward more sustainable solutions and adds to a broader movement toward sustainability.

Include Indoor Plants

Adding indoor plants to your house is a great method to improve the air quality and appearance of the space. Some plants are well known for their ability to filter pollutants from the air and enhance the general quality of the air indoors. Particularly good in eliminating pollutants are plants like peace lilies, spider plants, and snake plants.

Adding plants to your home also encourages wellbeing and relaxation. According to studies, having indoor plants can increase mood, attention, and lower stress. Adding plants to your home can improve its ambiance and lead to a healthier lifestyle.

Think about setting up a specific location for your indoor garden, whether it is a larger area with potted plants or a windowsill full of herbs. This will not only make your house look better, but it will also improve your general and mental wellness.

Encourage Community Engagement

Getting involved in your community can help you have more of an impact on the environment and strengthen your bonds with like-minded others. Take part in neighbourhood gardening projects, clean-up days, or environmental advocacy organizations in your community. Through these programs, you can learn from others about sustainable best practices and make a good impact on your community.

Participating in neighbourhood projects broadens your horizons and gives you the chance to exchange ideas and resources with like-minded people. You can encourage group action toward waste reduction and environmental stewardship by working with your community.

Additionally, try sponsoring courses or activities focusing on sustainability in your community. You may inspire people to adopt eco-friendly behaviours and create a network of support dedicated to building a sustainable future by sharing your knowledge and experiences.

To put it briefly, going green in your house is a multifaceted process that includes mindful consumption, waste minimization, indoor plant integration, water conservation, energy efficiency, and community involvement. You may greatly lessen your environmental impact and make your home healthier by putting these techniques into practice. This dedication not only helps the environment but also encourages others to follow suit, creating a sustainable culture that has a positive impact on our communities and beyond. We can create a more sustainable future for future generations and ourselves if we work together.

CHAPTER FOUR

Eco-Friendly Clothing And Lifestyle Options

Eco-Friendly clothing and lifestyle are an intentional effort to lessen our everyday influence on the environment while advancing moral behaviour and social justice. Fast fashion and materialism have become the standard in this day and age, thus this change highlights how important it is to make deliberate decisions that help the environment and all living things. In order to comprehend sustainable fashion and lifestyle choices, one must examine their importance, guiding ideals, and real-world implementations.

Comprehending Eco-Friendly Clothing

Fundamentally, sustainable fashion refers to apparel, accessories, and footwear made with a focus on reducing environmental damage and promoting moral labour standards. It takes a comprehensive strategy that takes into account every stage of a product's lifecycle, from the extraction of raw materials to their manufacturing, distribution, usage, and final disposal. Sustainable fashion's guiding principles emphasize the necessity of conscientious consumption and production practices that put the needs of people and the environment first.

A cornerstone of sustainable fashion is ethical production and sourcing. This idea emphasizes how crucial it is to select materials that are produced or harvested in a way that does not harm workers or deplete the environment. Brands committed to sustainability frequently choose organic, recycled, or up cycled fabrics that require less water, energy, and chemical inputs compared to conventional textiles. For example, less pollution is found in the soil and water when organic cotton is farmed without the use of artificial fertilizers or pesticides.

Furthermore, eco-friendly materials are promoted by sustainable fashion, which is important for minimizing environmental impact. This comprises materials with reduced carbon footprints and sustainable origins, like Tencel, hemp, and linen. Furthermore, a lot of manufacturers are reducing waste and the demand for new raw materials by including recycled materials into their collections. One example of this is recycled polyester created from plastic bottles.

The emphasis that sustainable fashion places on quality and durability is crucial. Sustainable fashion encourages buying high-quality clothing that will last a lifetime, as opposed to fostering a disposable society. Customers can drastically cut down on the frequency of purchases by opting for well-made apparel that they can use for many years, which will reduce waste and consumption in general.

Sustainable fashion techniques are further enhanced by the notion of cyclical fashion. When things are designed with their end-of-life in mind, they can be reused, recycled, or up cycled once they are no longer needed. This is known as circular fashion. This concept not only minimizes the quantity of textile waste that is disposed of in landfills, but it also promotes a sustainable culture in which goods are constantly repurposed within the economy.

Essentially, conscious consumption and a move toward minimalism are promoted by sustainable fashion. Focusing on buying only what is required and choosing classic items that may be combined, customers can lessen impulsive purchase behaviours that worsen the environment.

Realistic Sustainable Clothing Options

People can begin by investigating thrift and vintage stores as a means of embracing sustainable fashion practices. Unique pieces can be found in thrift stores, consignment stores, and online vintage clothing platforms, which also significantly lower the demand for new manufacture. This approach supports a more sustainable fashion environment in addition to assisting in the search for unique things.

Purchasing high-quality clothing is yet another essential sustainable fashion approach. Investing in classic, adaptable clothing that can be worn for a variety of seasons and events is advised instead of following fashions that promote frequent wardrobe turnover. Brands that put an emphasis on quality and longevity guarantee that the products you buy will stay fashionable and useful for many years.

Encouraging eco-friendly brands is crucial to the shift toward ethical fashion. Today, a lot of businesses use ethical business practices and open supply chains to showcase their sustainability initiatives, which helps customers make more informed decisions. Customers may push the fashion industry to embrace more sustainable practices by intentionally choosing to support these firms.

Exchanging clothes with others can be a fun and inventive approach to prolong the life of apparel. Through the organization of local garment exchanges or participation in them, people can update their wardrobes without having to buy new things. This encourages creativity and community ties in addition to sustainability.

One further empowering habit people might take up is learning to fix and upcycle garments. Learning the basics of sewing might help one make new clothes or mend old ones rather than throwing them away. Upcycling garments keeps them out of landfills, which not only fosters individual creativity but also decreases waste.

Realistic Sustainable Clothing Options

People can begin by investigating thrift and vintage stores as a means of embracing sustainable fashion practices. Unique pieces can be found in thrift stores, consignment stores, and online vintage clothing platforms, which also significantly lower the demand for new manufacture. This approach supports a more sustainable fashion environment in addition to assisting in the search for unique things.

Purchasing high-quality clothing is yet another essential sustainable fashion approach. Investing in classic, adaptable clothing that can be worn for a variety of seasons and events is advised instead of following fashions that promote frequent wardrobe turnover. Brands that put an emphasis on quality and longevity guarantee that the products you buy will stay fashionable and useful for many years.

Encouraging eco-friendly brands is crucial to the shift toward ethical fashion. Today, a lot of businesses use ethical business practices and open supply chains to showcase their sustainability initiatives, which helps customers make more informed decisions. Customers may push the fashion industry to embrace more sustainable practices by intentionally choosing to support these firms.

Exchanging clothes with others can be a fun and inventive approach to prolong the life of apparel. Through the organization of local garment exchanges or participation in them, people can update their wardrobes without having to buy new things. This encourages creativity and community ties in addition to sustainability.

One further empowering habit people might take up is learning to fix and upcycle garments. Learning the basics of sewing might help one make new clothes or mend old ones rather than throwing them away. Upcycling garments keeps them out of landfills, which not only fosters individual creativity but also decreases waste.

63 | Susan Ribble

CHAPTER FIVE

Realistic Sustainable Clothing Options

People can begin by investigating thrift and vintage stores as a means of embracing sustainable fashion practices. Unique pieces can be found in thrift stores, consignment stores, and online vintage clothing platforms, which also significantly lower the demand for new manufacture. This approach supports a more sustainable fashion environment in addition to assisting in the search for unique things.

Purchasing high-quality clothing is yet another essential sustainable fashion approach. Investing in classic, adaptable clothing that can be worn for a variety of seasons and events is advised instead of following fashions that promote frequent wardrobe turnover. Brands that put an emphasis on quality and longevity guarantee that the products you buy will stay fashionable and useful for many years.

Encouraging eco-friendly brands is crucial to the shift toward ethical fashion. Today, a lot of businesses use ethical business practices and open supply chains to showcase their sustainability initiatives, which helps customers make more informed decisions. Customers may push the fashion industry to embrace more sustainable practices by intentionally choosing to support these firms.

Exchanging clothes with others can be a fun and inventive approach to prolong the life of apparel. Through the organization of local garment exchanges or participation in them, people can update their wardrobes without having to buy new things. This encourages creativity and community ties in addition to sustainability.

One further empowering habit people might take up is learning to fix and upcycle garments. Learning the basics of sewing might help one make new clothes or mend old ones rather than throwing them away. Upcycling garments keeps them out of landfills, which not only fosters individual creativity but also decreases waste.

Hydropower: Using the flow of water, usually from rivers or dams, hydropower produces electricity. It can offer a steady and dependable power supply and is one of the earliest types of renewable energy. However, the environmental impact of dam construction and changes to natural streams must be considered.

Geothermal Energy: This type of energy uses the heat that naturally exists within the Earth to produce electricity or heat directly. This energy source works especially well in areas where there is a lot of volcanic activity. Geothermal systems can provide a continuous energy supply with minimum emissions.

Biomass Energy: Organic materials like plant and animal waste are used to make biomass energy. It can be utilized to produce energy, heat, and bio fuels. Although biomass has the potential to be a sustainable energy source, it must be managed carefully to prevent deforestation and other negative ecological effects.

Benefits of Green Energy

Making the switch to green energy has many advantages for people and society at large. Primarily, renewable energy sources contribute to the decrease of greenhouse gas emissions, which is a primary driver of climate change. We can lessen the effects that fossil fuel extraction and use have on the environment by reducing our dependency on them.

Green energy also encourages energy independence. Due to their heavy reliance on imported fossil fuels, many nations are susceptible to changes in price and geopolitical unrest. Investing in renewable energy can help countries become less reliant on foreign oil and become more energy secure.

Additionally, green energy promotes economic expansion and the development of jobs. It has been shown that the renewable energy industry provides a sizable number of jobs, from installation and manufacture to maintenance and research. The growth of the sector boosts local economies by generating possibilities for workers in a variety of fields.

Additionally, switching to green energy may result in long-term financial savings. The long-term advantages of lower electricity bills, fewer maintenance costs, and government subsidies frequently surpass the initial costs of investing in renewable energy systems, even though their initial cost may be higher than that of traditional sources. For example, solar panels have been shown by many businesses and homeowners to dramatically lower their energy bills over time.

How to Include Renewable Energy in Your Daily Life

Integrating green energy into your life can be accomplished in various ways, depending on your individual circumstances and preferences.

Install Solar Panels: One of the most effective ways to harness green energy at home is by installing solar panels. Homeowners can choose to install rooftop solar systems or participate in community solar programs, allowing them to access solar energy without needing to install panels on their property.

Switch to Renewable Energy Suppliers: Many utility companies now offer green energy options, allowing consumers to choose electricity sourced from renewable sources. By opting for these programs, individuals can support the growth of green energy while reducing their carbon footprint.

Utilize Energy Efficiency Practices: Enhancing energy efficiency in your home complements the use of green energy. Simple actions, such as using energy-efficient appliances, sealing windows and doors, and improving insulation, can help reduce overall energy consumption.

Consider Electric Vehicles: Electric vehicles (EVs) produce no tailpipe emissions and can be powered by renewable energy. Transitioning from a traditional gas-powered vehicle to an electric car not only reduces greenhouse gas emissions but also decreases reliance on fossil fuels.

Advocate for Policy Change: Supporting policies that promote renewable energy development can have a significant impact on the transition to green energy. Engaging with local representatives, participating in community discussions, and advocating for sustainable practices can help create a broader shift toward renewable energy.

Educate Others: Sharing information about the benefits of green energy and encouraging friends and family to consider sustainable practices can amplify the impact. Awareness and education are key to fostering a collective commitment to sustainability.

Green energy represents a pivotal shift toward a more sustainable and resilient future. By embracing renewable energy sources and making conscious lifestyle choices, individuals can significantly reduce their environmental impact while contributing to economic growth and energy independence. The benefits of green energy extend beyond individual choices, fostering a collective movement toward a healthier planet.

As the world continues to confront the challenges posed by climate change, the transition to green energy becomes increasingly vital. By understanding its principles and embracing practical solutions, we can power our lives sustainably and leave a positive legacy for future generations. Embracing green energy is not just a personal choice; it is a collective responsibility to protect our planet and ensure a viable future for all.

CHAPTER SIX

Eco-Friendly Travel And Transportation

The significance of environmentally friendly travel and transportation is growing along with environmental consciousness. The tourist business profoundly effects the environment, from carbon emissions linked with flights to the resources consumed by hotels and activities. Travellers can, however, reduce their environmental impact while still having a great time on their excursions by making thoughtful decisions and using sustainable methods. This chapter examines the fundamentals of environmentally responsible travel, viable modes of transportation, and helpful advice for choosing more sustainably when on the go.

The goal of eco-friendly travel is to promote sustainable practices while reducing the detrimental effects that tourism has on the environment and society. It entails making decisions that uphold local communities, safeguard the environment, and conserve cultural heritage. In order to ensure that future generations can benefit from the same experiences, the objective is to depart a place better than when you arrived.

Carbon emissions are a major issue with traditional travel. Particularly, air travel adds a substantial amount to greenhouse gas emissions. Individuals can lessen their carbon footprint and support a more sustainable tourist sector by implementing eco-friendly travel practices. This entails picking eco-friendly travel places, encouraging eco-friendly lodging, and taking part in ethical tourism practices.

Respecting local customs and communities is another aspect of sustainable travel. This entails interacting with locals, getting to know their traditions, and patronizing neighbourhood businesses. Travellers can bolster the local economy and learn more about the area by opting to shop at local markets, eat at neighbourhood eateries, and take part in community-led excursions.

Greener Options for Transportation

Transportation is a vital part of eco-friendly travel. Travellers may want to take into account the following sustainable options:

Public Transport: Making use of public transportation, such trains, buses, and trams, is one of the best methods to lessen your carbon footprint. Travellers who use public transit can experience a place like a native and save a lot of energy compared to driving alone.

Biking: A healthy and environmentally beneficial way to explore is by bike, and many cities have bike-sharing programs. Biking saves pollution, gives a fantastic way to see the sights, and often allows travellers to reach areas that may be inaccessible by car.

Walking: For shorter distances, walking is the most sustainable alternative. It enables visitors to become fully immersed in their surroundings, find hidden treasures, and establish a connection with the community.

Carpooling and Ridesharing: If public transit is not an option, think about carpooling or using environmentally friendly ridesharing services. As a result, there are fewer cars on the road and fewer pollutants.

Electric Vehicles: If you must rent a car, you can greatly reduce your environmental effect by choosing an electric or hybrid vehicle. Travellers can now select sustainable modes of transportation more easily because a large number of automobile rental firms provide eco-friendly options.

Rail Travel: Compared to driving or flying, rail travel can be a more environmentally friendly option in areas where trains are available. Trains typically emit fewer emissions per passenger compared to vehicles and airlines, making them a cleaner choice for longer journeys

Tips for Eco-Friendly Travel

In addition to choosing sustainable transportation, travellers can adopt various eco-friendly practices throughout their journeys:

Choose Eco-Conscious Accommodations: Look for hotels and lodgings that prioritize sustainability through energy-efficient practices, water conservation, and support for local communities. Many eco-friendly accommodations have certifications, such as Green Key or Earth-Check, that indicate their commitment to sustainability.

Pack Light: Travelling with less luggage not only makes it easier to move around but also reduces carbon emissions associated with transportation. The heavier the luggage, the more fuel is needed for flights and vehicles. Aim to pack only what you need and consider using eco-friendly travel products, such as reusable toiletries.

Reduce Single-Use Plastics: Bring reusable water bottles, shopping bags, and utensils to minimize single-use plastic waste during your travels. Many destinations have refill stations or water fountains where you can refill your bottle, reducing the need for bottled water.

Support Local Businesses: Eating at local restaurants, shopping at markets, and using local guides helps support the economy of the destination. It also ensures that your money stays within the community rather than going to large corporations.

Participate in Eco-Friendly Activities: Choose activities that promote sustainability, such as guided nature walks, wildlife conservation programs, or volunteering opportunities. Engaging in responsible tourism practices helps protect the environment and supports local conservation efforts.

Leave No Trace: When exploring natural areas, follow the Leave No Trace principles. This entails clearing up any waste, adhering to established routes, and showing consideration for wildlife. Leaving natural areas undisturbed helps preserve their beauty for future visitors.

Educate Yourself: Before travelling to a new destination, learn about its environmental challenges and cultural customs. Understanding the local context helps you make informed decisions that align with sustainable practices.

Eco-friendly travel and transportation represent a critical shift toward more responsible and sustainable tourism practices. By making conscious choices about how we travel, where we stay, and what activities we engage in, we can significantly reduce our environmental footprint and positively impact the communities we visit.

As global citizens, we have the power to advocate for sustainable travel practices that benefit both the planet and local populations. By embracing eco-friendly travel, we contribute to a movement that prioritizes preservation, respect, and sustainability, ensuring that future generations can explore and enjoy the wonders of our world. Whether it's a weekend getaway or a long journey, adopting eco-friendly practices in our travel habits is essential for a sustainable future.

CHAPTER SEVEN

The Importance Of Ethical Consumption

The trend toward ethical consumption is largely driven by consumers. People may encourage firms to embrace more egalitarian and sustainable practices by making educated decisions. This necessitates a dedication to education because customers have to do the legwork to learn about businesses and their policies, including looking up product origins, labour practices, and environmental regulations.

Consumer voices have been amplified by social media and technology, enabling people to exchange knowledge and promote moral behaviour. Platforms dedicated to ethical buying and product reviews empower individuals to make better educated decisions, producing a ripple effect that drives brands to be more accountable and transparent.

At its core, ethical consumption is being aware of the broader effects of the things that one chooses to buy. It means considering how products affect society and the environment in addition to the processes involved in their production, delivery, and disposal. This covers a wide range of topics, such as how products are treated for animals, how labour is done, how resources are exploited, and how products affect the environment. Customers may support companies that share their values and contribute to the creation of a more sustainable and equitable economy by prioritizing products that are created and sourced ethically.

A few instances of ethical consumerism include choosing to buy fair-trade products, supporting local businesses, reducing waste through minimalism, and making ecologically responsible purchases. Consumers are often required to look for more details about the products and brands they choose, as well as transparency and accountability from companies. Through cultivating a more morally-minded consumer culture, people can work together to impact markets and create positive change.

The Significance of Ethical Consumption

One of the main reasons for ethical consumerism is to advance social fairness. Many goods, especially those in the food and fashion sectors, are made in exploitative environments where people are mistreated, overworked, and denied fundamental rights. Customers may support fair salaries and improved working conditions for labourers worldwide by purchasing things made responsibly.

The decisions made by consumers have a significant impact on the environment. Pollution, resource depletion, and climate change are largely caused by fast fashion, excessive plastic use, and unsustainable agriculture. Ethical consumerism encourages individuals to purchase products that limit harm to the environment, such as those created from sustainable resources, with eco-friendly packaging, or those that promote conservation measures.

In addition, ethical consumption generally means supporting local enterprises over global ones. By assisting regional producers, we can build stronger communities, create jobs, and promote a feeling of community. Buying locally helps create a more resilient economy that prioritizes sustainability and the welfare of the community.

Additionally, ethical consumer behaviour forces businesses to take on more accountable roles. Supporting companies who are transparent about their sourcing, labour practices, and environmental effect will encourage others to do the same, which will ultimately result in a more accountable marketplace in a time when consumers are requesting more openness from brands.

The treatment of animals is included in the concept of ethical consumption. Consumer awareness of the brutality involved in factory farming and animal testing for products is growing. Customers may support animal rights and promote more compassionate treatment across businesses by purchasing cruelty-free and compassionate products.

Moreover, many ethical consumption habits value products that are created using traditional methods or that assist indigenous cultures. Customers can help local customs and crafts that may be in danger from globalization by making the decision to buy products that respect and maintain cultural heritage.

The Role of Consumers in Driving Change

The trend toward ethical consumption is largely driven by consumers. People may encourage firms to embrace more egalitarian and sustainable practices by making educated decisions. This necessitates a dedication to education because customers have to do the legwork to learn about businesses and their policies, including looking up product origins, labour practices, and environmental regulations.

Consumer voices have been amplified by social media and technology, enabling people to exchange knowledge and promote moral behaviour. People are empowered to make better decisions by platforms devoted to ethical buying and product reviews, which in turn causes firms to become more transparent and accountable.

Collective action can also bring about a great deal of change. The ability of consumer voices to influence market trends has been demonstrated by the successful pressure that consumer campaigns and grassroots movements have had on businesses to adopt ethical practices.

It is impossible to exaggerate the significance of ethical consumption. People can effect change through their shopping decisions in a world where every purchase has an impact on environmental sustainability, animal welfare, and social justice. By prioritizing ethical consumption, consumers may promote fair labour practices, safeguard the environment, support local economies, and fight for the humane treatment of animals.

It is critical for consumers to be aware and involved as the discourse surrounding ethical consumerism develops. Individual decisions can have a tremendous aggregate impact on industry, resulting in a more equitable and sustainable world. In the end, ethical consumption honours our interdependence and our duty to the environment and to one another. It is more than just a fad. By practicing mindful consumption, people may help ensure that everyone has a brighter future.

CHAPTER EIGHT

Long-Term Habits for Sustainable Living

Making the switch to sustainable living is crucial to protecting the ecosystem and leaving the earth healthy for coming generations. While taking action right away might have a good effect, long-term habit development is essential for bringing about change that lasts. Living sustainably entails making deliberate decisions that lessen our impact on the environment, advance social justice, and promote prudent resource management.

Aware Consumption

Adopting conscientious consumption is one of the long-term habits with the biggest influence. Making deliberate selections about what to buy and from whom is part of this habit. Quality above quantity is a great way for people to cut waste and support ethical brands. This entails making investments in long-lasting goods, encouraging regional enterprises, and looking for businesses that use sustainable techniques. To make sure that their purchases are consistent with their ideals, conscientious consumers frequently investigate the sources of products, labour methods, and environmental impact. With time, this behaviour reduces individual consumption and pushes businesses to use more environmentally friendly procedures.

Reducing Waste

Another essential habit for maintaining a sustainable lifestyle is cultivating a mindset of waste minimization. Reducing, reusing, and recycling are all part of this. Assess daily routines and pinpoint places where waste can be cut first. For example, using reusable containers, bags, and utensils can dramatically reduce the amount of waste produced by single-use plastics. Furthermore, composting organic waste—such as leftover fruit and vegetable scraps—can improve soil quality and reduce landfill inputs. Through persistent efforts to reduce waste, people can cultivate a sustainable culture within their homes and communities.

Ecological Automobile

Long-term sustainable living entails reconsidering transportation habits. Choose environmentally responsible modes of transportation whenever you can. Short travels by bike or foot not only save carbon emissions but also improve physical health. When travelling longer distances, think about taking the public transportation, carpooling, or travelling in an electric or fuel-efficient car. People may drastically minimize their carbon footprints, help create cleaner air, and ease traffic congestion by giving priority to sustainable transportation options.

Water Conservation

Water conservation is a crucial—yet usually ignored—aspect of sustainable living. Changing one's behaviour to use less water over time can save a lot of money. Simple solutions include repairing leaks as soon as possible, using water-saving devices, and collecting rainwater for gardening. Moreover, cutting back on the quantity of water needed for daily chores like brushing your teeth and cleaning dishes will help save this vital resource. By incorporating water-saving habits into their daily lives, individuals may encourage sustainability and preserve neighbouring ecosystems.

Encouraging Sustainable and Local Agriculture

A another long-term, beneficial practice is to support sustainable, local agriculture. Choosing to purchase locally grown vegetables helps to build community, supports local farmers, and lessens the carbon footprint involved with transportation. Participating in farmers' markets and community-supported agriculture (CSA) initiatives can strengthen the bonds between producers and customers, promoting more resilient and sustainable food systems. Growing food on one's own, even in a tiny space, can also increase self-sufficiency and lessen dependency on industrial agriculture.

Accepting Simplicity

A simpler lifestyle that prioritizes quality over quantity is encouraged by minimalism. People can simplify their life and lessen their need for material possessions by taking a minimalist approach. This behaviour promotes thoughtful shopping and assists people in putting more emphasis on experiences than material belongings. By prioritizing experiences over stuff, people can cultivate gratitude, alleviate stress, and foster deeper relationships with their communities and the environment.

Ongoing Education and Protest

Finally, long-term transformation requires a commitment to ongoing education about sustainability challenges. Making educated decisions can be facilitated by keeping up with new sustainable practices, environmental issues, and developing technologies. Advocacy is also essential; people may make a bigger difference by taking part in local projects, endorsing laws that advance sustainability, and motivating others to follow sustainable lifestyles. Talking about sustainability encourages awareness and accountability, which eventually results in a more sustainable future.

Setting up long-term routines for sustainable living calls for commitment, awareness, and flexibility. People can significantly improve their communities and the environment by adopting sustainable transportation, minimizing waste, emphasizing energy efficiency, adopting conscious consumerism, conserving water, supporting local agriculture, embracing minimalism, and committing to ongoing education and advocacy.

These behaviours support a more just and equitable society in addition to a healthier world. By committing to these behaviours, people join a broader movement towards sustainability and promote social responsibility and environmental care in society. In the end, the path to sustainable living is an ongoing one, and each little action made now can result in a more sustainable and rewarding future.

CONCLUSION

It is evident as we close ECO LIVING MADE SIMPLE: A Practical Guide to Sustainable Choices for Modern Life that adopting a sustainable lifestyle is essential for the health of our planet and future generations, not merely a fad. We have examined the various ways that thoughtful, little decisions can have a big beneficial influence on our communities, our environment, and ourselves along this journey.

Living sustainably encourages us to reassess our routines and make deliberate decisions that are consistent with our ideals rather than calling for significant adjustments or sacrifices. Every action we take can help create a society that is healthier and more sustainable, from embracing conscious consumption and supporting local businesses to cutting back on waste and choosing eco-friendly items.

Always keep in mind that every little step matters when you set out on your own eco-living adventure. Create a group of like-minded people who are dedicated to changing the world by imparting your expertise, motivating others, and building community. When we work together, we can make a bigger impact than we could if we worked alone.

This is not where the journey ends. Sustainability is a way of life that changes as we gain knowledge and experience. Accept the difficulties, acknowledge your accomplishments, and keep an open mind to fresh concepts and methods that will improve your eco-living experience.

Let us decide to be the change we want to see as we navigate a world that is always changing. By making sustainability a priority in our day-to-day activities, we not only safeguard the environment but also improve our own lives by giving them meaning, fulfilment, and a connection to the natural world.

We appreciate your participation as we move toward a more sustainable future. One small decision at a time, here's to a better, more environmentally friendly tomorrow.